A. K. Baikulov
O. S. Tashanov

TOXICOLOGICAL CHEMISTRY

A. K. Baikulov
O. S. Tashanov

TOXICOLOGICAL CHEMISTRY

**Pharmacy education courses: behind 60910700
WORKSHEET IV course**

ScienciaScripts

Publisher:
Sciencia Scripts
is a trademark of
Dodo Books Indian Ocean Ltd. and OmniScriptum S.R.L publishing group

120 High Road, East Finchley, London, N2 9ED, United Kingdom
Str. Armeneasca 28/1, office 1, Chisinau MD-2012, Republic of Moldova, Europe
Managing Directors: Ieva Konstantinova, Victoria Ursu
info@omniscriptum.com

Printed at: see last page
ISBN: 978-620-8-15767-8

Contents

SAMARKAND STATE MEDICAL UNIVERSITY
PHARMACEUTICAL AND TOXICOLOGICAL CHEMISTRY DEPARTMENT

Pharmacy education courses

WORKSHEET

IV course

CASE STUDIES

Science: "Toxicological Chemistry".

(Total allotted time - 144 hours)

Topics of practical classes		Clock
IV course		
1	Introduction to toxicological chemistry. Facilities and laboratory equipment. Requirements for.	4
2	Understanding physical evidence. Preliminary examination of physical evidence: pH-environment, colour, odour and preliminary examination of the object for the presence of individual substances (acids, alkalis, cyanides, white arsenic).	4
3	Removal of toxic substances from the site by steaming. Examination of the first distillate for cyanidic acid	4
4	Analysis of volatile substances in the distillate for formaldehyde and acetone.	4
5	Analysis of acetic acid and phenol by volatiles in the distillate	4
6	Examination of the second distillate for chloroform, chloral hydrate, carbon tetrachloride and their differentiation.	4
7	Chemical analysis of methyl, ethyl, amyl alcohols from distillate.	4
8	Introduction to gas-liquid chromatographs. Definition quality and quantity of alcohol in blood and urine by GSX method.	4
9	Wet mineralisation of the property in the presence of sulphuric and nitric acid acids. Denitrification of mineralisate and preparation for analyses	4
10	Determination of barium and lead cations after separation of the precipitate from the mineralisate.	4
11	Determination of manganese, chromium, cations in mineralisate.	4
12	Determination of copper and silver cations in mineralisation	4
1 7 13	Determination of zinc, antimony, thallium and cadmium cations in the mineralisate	л 4
14	Determination of bismuth and aluminium cations in mineralizate.	4
15	Determination of cobalt and tin cations in mineralisate.	4

1 A 16	Determination of arsenic cation in mineralisation . Analysis of an unknown object.	л	4
1 7 17	Liver and kidney damage. Determining the quality and quantity of mercury by degradation.		
18	Forensic Chemistry. Forensic chemists, their responsibilities, ethics and deontology. Writing and defence of expert evidence conclusions		4
19	Extraction of toxic substances from biological objects with using polar solvents. Separation methods with using the acidified water method.		4
on 20	Liquid-liquid extraction. Determination of phenacetin, salicylic acid, acetylsalicylic acid in the extract.	л	4
7 1 21	Phenazone (antipyrine), propiphenazone (amidopyrine), metamizole sodium (analgin), phenylbutazone (butadione) from excreta and liquids determination of substances by the YuQX method.	л	4
22	Preparation of reagents used in the analysis of alkaloids. Determination of indole alkaloids strychnine, brucine, Reserpine in separators and liquid materials.		4
23	Determination of purine alkaloids (caffeine, theobramine, theophylline) in extracts and liquid materials. Caffeine Dependence. Spectrophotometric analysis.		4
24	Determination of the alkaloids nicotine, anabasine and paxicarpine in extract and liquid material. Tobacco addiction. Microcrystallographic analysis.		4
25	Determination of the alkaloids atropine, hyoscyamine, scopolamine in the extracts and liquid materials. Applications of gel chromatography	л	4
26	Determination of quinine and papaverine alkaloids in extracts and liquid materials. Applications of gel chromatography.		
77 2 7	Identification of novocaine, dicaine, dimedrol, lidocaine substances in secretions and fluid samples. FEC analysis.	л	4
28	Determination of dimedrol, amitriptyline and depressants in secretions and liquid substances. Analysis by TDSIS		4
29	Diuretic agents in forensic toxicological practice. We analyse them using the YuQX screening method.		4
30	Anti-inflammatory drugs in forensic medicine toxicology. We analyse them in the YuSSX method.		4
31	Antidiabetic drugs in forensic toxicology		4

	practice. We analyse them using the YuQX screening method.	
32	Methane and carbon monoxide poisoning. Analysis blood carboxyhaemoglobin by chemical and physicochemical	4
33	Dialysis. Poisons that can be isolated from an object using water. Extraction of mineral acids, alkalis and their salts and determination from dialysate. Mineral acids (sulphuric acid, hydrochloric acid, nitric acid).	4
34	Agricultural pesticides, organophosphorus pesticides, organochlorine toxic chemicals. Separation and identification chlorophos, carbophos, hexachlorane.	4
35	Derivatives of synthetic pyrethroids. Cypermethrin, Danitol, Decis. Methods of extracting them from objects and analysing them. Writing and defence of expert reports.	4
36	Separately analysed substances. Identification of an unknown object. Writing and defence of expert reports.	4
General		**144**

Exercise #1 .

THEME: Introduction to toxicological chemistry. Premises and laboratory equipment. Requirements for them.

Laboratory training plan

Familiarity with toxicological chemistry, subjects and laboratory equipment.

2. They to be accommodated _ requirements

Training will be conducted : For students of toxicological chemistry analysis for sent objects _ _ _ general condition , features _ _ and initial with t h e kshrisham dependent _ _ was _ _ conditions taught _ _ _ _

Expected outcomes during lab training:

- types of objects, their preparation for analysis;
- Determining the packaging and integrity of the items being shipped ;
- Examination of the environment of physical evidence, colour, consistency ; additions that can be found in the object, their division and other cases are introduced.

Required equipment for laboratory classes : Textbook, lecture text, teaching aids, funnel, test tubes, reagents, demonstration materials, tape, paper.

Laboratory Equipment : A room equipped with specialised laboratory equipment.

TASK 1:

Sampling of biological objects and their preparation for analysis

CHALLENGE 2:

A selection of objects in the package.

CHALLENGE 3:

Preparation of the biological object (liver) for analysis

Questions to test knowledge

1. Significance and objectives of toxicological chemistry.
2. Toxicological chemical facilities and their requirements.
3. Check the integrity and morphological structure of the object .
4. Determining the status of physical evidence.

Date

Exercise #2.

THEME: The concept of physical evidence.

Preliminary examination of physical evidence: pH-environment, colour, smell and preliminary examination of the object for the presence of individual substances (acids, alkalis, cyanides, white arsenic).

Lab workout plan

1. Introduction to toxicological chemistry, understanding of laboratory equipment, physical evidence.

2. Preliminary examination of physical evidence

Training will be conducted : For students toxicological chemistry analysis for sent objects _ _ _ general condition , features _ _ and initial t h e xhrish with dependent _ _ was _ _ laryngeal conditions teach _ _ _ _

Expected outcomes during lab training:

- physical evidence;
- Determining the packaging and integrity of the items being shipped ;
- Examination of the environment of physical evidence, colour, consistency ; additions that can be found in the object, their division and other cases are introduced.

Necessary equipment for laboratory classes :

Textbook, lecture text, teaching aids, water vapour drying apparatus, funnel, test tubes, reagents, visual materials, scotch tape, paper

Laboratory Equipment : A room equipped with specialised laboratory equipment.

1- CHALLENGE.

Preliminary examination of physical evidence *Determination of the appearance of the object being inspected.*

TASK 2 consistency and morphological structure of a biological object
Determine whether an object has colour, odour, foreign substances or their residues .

CHALLENGE 3

the reaction of the object sent for analysis to various parameters
Conduct a preliminary chemical study before starting the main full analysis.

CHALLENGE 4

the reactions of the object to be analysed to various parameters. Before starting the main full analysis, carry out a preliminary chemical test.

Questions to test knowledge

1. Significance and objectives of toxicological chemistry.

(2) Determining the status of physical evidence.

It's not raining. Date

Exercise #3
Removal of toxic substances from a facility using water vapour . First distillate acid cyanide check.

Laboratory training plan

1. Separation of "volatile" poisons from the composition of a biological object using water vapour.Theoretical basis .

2. Complete assembly of the Water Vapour device.

3. Extraction of volatile substances from the facility.

4. Determination of cyanic acid in distillates of substances extracted from a site using the water bug .

Exercise _ _ Objective : Students should be able to plan a forensic chemical analysis

Forensic chemical analysis and practice the theoretical basis of extracting "volatile" poisons from the composition of a biological object using water vapour, present the device used in this process, fully assemble the device and remove volatile substances, learn to separate from the object and determine cyanide acid from distillates.

During laboratory sessions Expected Outcomes:

-   Assembly of the drive apparatus using water vapour;

-   Preparation of a biological object for movement using water vapour ;

-   "Pilots" learn how to extract toxic substances using water vapour and analyse them using chemical, pharmacological and physicochemical methods;

Laboratory teaching _ _ transfer _ _ for necessary tools - equipment: Textbook , lecture text , reading _ _ instructions , Water _ _ boogie using driving apparatus , funnel , test tubes , reagents , exhibition materials , tape , paper _ _ _

Laboratory equipment: Room equipped with specialised laboratory equipment .

CHALLENGE 1

Cyanide acid in determining Berlin _ _ blue crop do from reaction _ using determine _

CHALLENGE 2

Transfer of "volatile" poisons into the human body based on theoretical knowledge of toxicological chemistry;

-   distribution in the human body,

-   you're eating me,

-   explain the tooth that comes out of the body.

Toxicological significance of cyanic acid, its uses, symptoms of poisoning and metabolism .

reaction is used to determine the authenticity of cyanoic acid ?

Date

Exercise #4_ _ _ _ _
Analysis for formaldehyde and acetone from volatile substances in the
distillate .

Laboratory training plan

1. Checking the second distillate for formaldehyde and acetone .
2. Determination of formaldehyde substances in distillates of substances extracted from a facility using an aqueous tourniquet .
3. Checking the second distillate for phenol.
4. Determination of acetic acid in the distillate of substances extracted from a facility using an aqueous tourniquet .

Training Objective : To familiarise students with the toxicological significance of acetone and formaldehyde in second distillate and the forensic chemical reactions used to determine the authenticity of .

Expected outcomes during lab training:

-   Determination of acetone in the distillate of substances extracted from a facility using water vapour .
-   The formaldehyde content of the second distillate was determined .
-   Determination of phenol in the distillate of substances extracted from a facility using water vapour .
-   Acetic acids in the second distillate were determined .

Necessary equipment for laboratory classes :

Textbook, lecture text, teaching aids, water vapour drying apparatus, funnel, test tubes, reagents, visual materials, tape, paper.

Laboratory Equipment : A room equipped with specialised laboratory equipment.

CHALLENGE 1

formal e gid
The reaction is carried out with an alkaline solution of resorcinol .

CHALLENGE 2

Code in reaction with an alkaloid.

CHALLENGE 3

Reaction of fuchsin with sulphitic acid .

4- CHALLENGE.

Reduction reaction of silver oxide .

Date

11

<h1 style="text-align:center">Exercise #5.</h1>

<h2 style="text-align:center">The volatiles in the distillate are analysed for acetic acid.
and phenol.</h2>

Laboratory training plan

1. Determination of formaldehyde substances in distillates of substances extracted from an object with water

2. Examination of the second distillate for phenol.

3. Determination of acetic acid in the distillate of substances extracted from an object with water

Training Objective : To familiarise students with the toxicological significance of acetone and formaldehyde in second distillate and the forensic chemical reactions used to determine the authenticity of .

Expected outcomes during lab training:

- Determination of acetone in the distillate of substances extracted from a facility using water vapour .

- The formaldehyde content of the second distillate was determined .

- Determination of phenol in the distillate of substances extracted from a facility using water vapour .

- The acetic acids in the second distillate were determined .

Necessary equipment for laboratory classes :

Textbook, lecture text, teaching aids, water vapour drying apparatus, funnel, test tubes, reagents, visual materials, tape, paper.

Laboratory Equipment : A room equipped with specialised laboratory equipment.

1 - CHALLENGE

ACETIC ACID - CH_3COOH

Determination of authenticity. *Reaction with ferric (III) chloride*

CHALLENGE 2
acetic acid ..

CHALLENGE 3
Zero matter formation reaction .

Date

Exercise #6
Examination of the second distillate for chloroform, chloral hydrate, carbon tetrachloride and their differentiation.

Laboratory training plan

1. Differentiation of volatile chlorine-preserving toxicants from each other .

Training objective: To teach students to distinguish toxicologically significant chlorine-containing preservatives in distillate **from each other**.

Expected outcomes during lab training:

- Conducts reactions to distinguish the substances chloroform, chloral hydrate, and carbon tetrachloride from each other.

Required equipment for laboratory classes : Distillate, test tubes, test tubes, reagents, demonstration materials, tape, paper.

Laboratory equipment: Room equipped with specialised laboratory equipment

.

CHALLENGE 1

Transition of organic bound chlorine to dissociable chlorine anion:
Reaction for the formation of isonitrile .

CHALLENGE 3

Reduction reaction of copper(P) hydroxide by phelene liquid .

Date

Exercise 7
Chemical analyses of methyl, ethyl, amyl alcohols from distillate .

Laboratory training plan

1. Chemical analysis of methyl alcohol from distillate .
2. Chemical analysis of ethyl alcohol from distillate .
3. Chemical analysis of amyl alcohol from distillate.

Course objectives: Students should have information about the toxicological and narcological significance of alcohols, their narcotic types, master the analysis of **distillates.** It is required to be able to find out the forensic chemical significance of the reactions used in determining the authenticity of these substances by means of chemical reactions.

Expected outcomes during lab training:

-   Chemical analysis of methyl alcohol from distillate .
-   Chemical analysis of ethyl alcohol from distillate .
-   Chemical analysis of amyl alcohol from distillate.

Necessary equipment for laboratory classes: Distillate, flasks, test tubes, reagents, demonstration materials, tape, paper.

Laboratory equipment: A room equipped with specialised laboratory equipment.

CHALLENGE 1

I was before alcohol.

Reaction of ester formation .

CHALLENGE 2

methyl alcohol to formaldehyde, then determination of the formaldehyde formed.

CHALLENGE 3

Ethyl alcohol *Iodoform reaction* . _

CHALLENGE 3

Amyl and isoamyl alcohols

***Reaction involving salicylic aldehyde and concentrated sulphuric acid* .**

Date

Exercise #8
Introduction to gas-liquid chromatographs.
Determination of the quality and quantity of alcohol in blood and urine
by the GSX method.

Laboratory training plan

1. Familiarity with the operating procedures of gas-liquid chromatographs.

2. Determination of alcohol in qualitative blood and urine by GSX method.

Learning objective: To familiarise students with the operation of gas-liquid chromatographs. It is required to measure the alcohol content in blood and urine using the GSX method.

Expected outcomes during lab training:

You are familiar with the operating procedures of gas-liquid chromatographs .
- Learns how to determine alcohol in blood and urine using the GSX method.

Necessary equipment for laboratory classes: Distillate, flasks, test tubes, reagents, demonstration materials, tape, paper.

Laboratory Equipment : A room equipped with specialised laboratory equipment.

CHALLENGE 1

Conduct gas chromatographic analysis

CHALLENGE 2

Some conditions for determining the quality of alcohol

CHALLENGE 3

Determining the presence of ethyl alcohol in human blood and urine

Date

Exercise #9

Mineralisation of the site in the presence of sulphate and nitrate acids. Denitrification of the mineralisation and preparation for analyses .

Laboratory training plan

1. Mineralisation of biobjects with mixtures of sulphuric and nitric acids.
2. Denitrification of mineralisate and preparation for assays.

Learning objective: During laboratory sessions students will be introduced to the processes of mineralisation (decomposition) of heavy metal compounds from the composition of the biological object under study using sulphuric and nitric acids, as well as the processes of denitrification. and preparation for analyses.

Expected outcomes during lab training:

1. Take an object containing metal compounds during a laboratory activity and examine its appearance;
2. Identification of oxidising agents from the composition of mineralisate, and familiarisation with the method of their removal by chemical means, i.e. formaldehyde;
3. Preparation of denitrated mineralisate for determination of metal cations ;
4. Familiarisation with the method of extraction of heavy metal compounds from the

structure of the biological object under study by mineralisation (decomposition) using sulphuric and nitric acids;

5. After completing the work, the student records the conclusion in the workbook

The workbook after checking the correctness of the work results with the teacher.

Necessary equipment for laboratory classes: Keldahl flask, tripod, gas burner, aspiration cabinet, reagents, demonstration materials, tape, paper.

Laboratory Equipment : A room equipped with specialised laboratory equipment.

CHALLENGE 1

Mixtures of sulphuric and nitric acids of biological objects mineralisation with

---

CHALLENGE 2

Determination of oxidising agents in mineralisate.

CHALLENGE 3

What substance is used to determine the authenticity of silver in a painting?

What does the reaction in the picture mean?

$$2CuCl_2 + K_4[Fe(CN)_6] \quad \cdot \quad \downarrow Cu_2[Fe(CN)_6] + 4KCl$$

$$Cu_2[Fe(CN)_6] + 2CdCl_2 \quad \cdot \quad \downarrow 2CuCd[Fe(CN)_6] + 4KCl$$

4. Which manganese compound is shown in the figure?

Date

Exercise #10

Determination of barium and lead cations after separation of the precipitate from the mineralisation .

Laboratory training plan

1. Determination of barium cation in the mineralisation.

2 Determination of lead cation in the mineralisation.

Purpose of the lesson: to acquaint students with the toxicological significance and methods of analysis of some metallic poisons extracted from biological objects by mineralisation (decomposition) using sulphuric and nitric acids. In the laboratory, students learn the conditions for analysing pure barium solutions, lead preservatives and solutions of unknown composition.

Expected outcomes during lab training:

1.   Separation of the sludge contained in the mineralisation ;

2.   Separation of barium and lead sulphates from the precipitate;

3.   Determination of the purity of barium and lead cations;

4.   After identifying an unknown substance given by the teacher, they have to check the correctness of the definition of the substance according to the material features, learn to summarise in a practical notebook, write a conclusion and sign it with the teacher of the group.

Necessary equipment for laboratory classes: Mineralisate, filter paper, funnel, reagents, test tubes, demonstration materials, tape, paper.

Laboratory Equipment : A room equipped with specialised laboratory equipment.

CHALLENGE 1

Separation of barium and lead sulphate precipitates.

CHALLENGE 2

Recrystallisation reaction of barium sulphate .

Reaction of barium cation with potassium chromate and potassium dichromate

3 - CHALLENGE.

ANALYSIS OF CATION OF SWINZA action with dithizone

TASK 4 *precipitation of lead sulphate .*

CHALLENGE 5

Formation of potassium, lead and copper hexonite .

CHALLENGE 5

Zn $_s$ P $_{2+}$ 6HCl®

to complete the reaction?

Which coloured substance is formed by this reaction?

18

$$H_3PO_4 + 12(NH_4)_2MoO_4 + 21HNO_3 \rightarrow$$
$$(NH_4)_3PO_4 \cdot 12MoO_3 + 21NH_4NO_3 +$$
$$12H_2O$$

What substance and precipitate of what colour is formed in this reaction?

$$2Sb^{+3} + 3Na_2S_2O_3 + 3H_2O \rightarrow$$
$$Sb_2S_3 + 3Na_2SO_4 + 6H^+$$

Which method is used to determine the authenticity of substances?

$$Cd^{+2} + 2\ \underset{C_2H_5}{\overset{C_2H_5}{>}}N-C\overset{S}{\underset{SH}{<}} \ \xrightarrow[2{,}5\ n\ NaOH]{pH\text{-}12}\ \underset{C_2H_5}{\overset{C_2H_5}{>}}N-C\overset{S}{\underset{S}{<}}Cd/2 + 2H^+$$

$$\underset{C_2H_5}{\overset{C_2H_5}{>}}N-C\overset{S}{\underset{S}{<}}Cd/2 + 2HCl \longrightarrow CdCl_2 + 2\ \underset{C_2H_5}{\overset{C_2H_5}{>}}N-C\overset{S}{\underset{SH}{<}}$$

Exercise #11

Manganese and chromium from mineral cation determination.

Laboratory training plan

1. manganese in the mineralisation .

2. chromium in the mineralisation .

Purpose of the class: To acquaint students with the toxicological significance and methods of analysis of metal poisons, manganese, chromium and silver cations extracted from biological objects by mineralisation (decomposition) using sulphuric and nitric acids.

Expected outcomes during lab training:

1. determination from the filtrate of the manganese cation that may be present in the mineralisation;

2. chromium and silver that may be present in the mineralisation from the leachate;

Necessary equipment for laboratory classes: Mineralisate, filter paper, funnel, reagents, test tubes , demonstration materials, tape, paper.

Laboratory equipment: Room equipped with specialised laboratory equipment

.

CHALLENGE 1

Oxidation reaction of manganese cation by potassium periodate

TASK 2 *manganese* cation *by ammonium persulphate*

CHALLENGE 3

CHROMIUM CATION ANALYSIS

Reaction for the formation of nachromic acid.

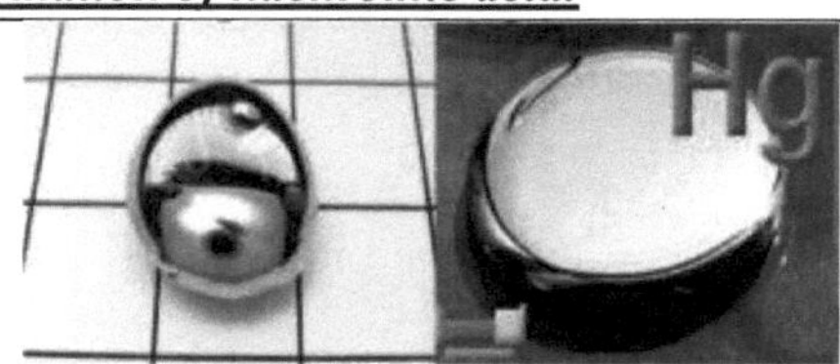

Date

Exercise 12
copper and silver mineralisation.

Laboratory training plan

3. of copper in the mineralisation.

4. silver in the mineralisation .

Purpose of the class: To acquaint students with the toxicological significance and methods of analysis of metal poisons, manganese, chromium and silver cations extracted from biological objects by mineralisation (decomposition) using sulphuric and nitric acids.

Expected outcomes during lab training:

3. Determination from the filtrate of the manganese cation that may be present in the mineralisation;

4. chromium and silver that may be present in the mineralisation from the leachate;

Necessary equipment for laboratory classes: Mineralisate, filter paper, funnel, reagents, test tubes, demonstration materials, tape, paper.

Laboratory equipment: Room equipped with specialised laboratory equipment

.

1   - CHALLENGE

Copper cation analysis

Reaction to form copper diethyldithiocarbaminate

CHALLENGE 2

Reaction to form copper and cadmium ferrocyanide .

CHALLENGE 3

Silver cation analysis

Reaction for the formation of silver dithionate compound .

CHALLENGE 4

2   *Silver chloride precipitate formation reaction .*

3   . To which class does this substance belong?

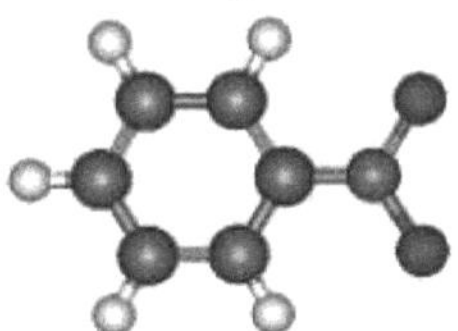

4. What substance is obtained in the laboratory by this method?

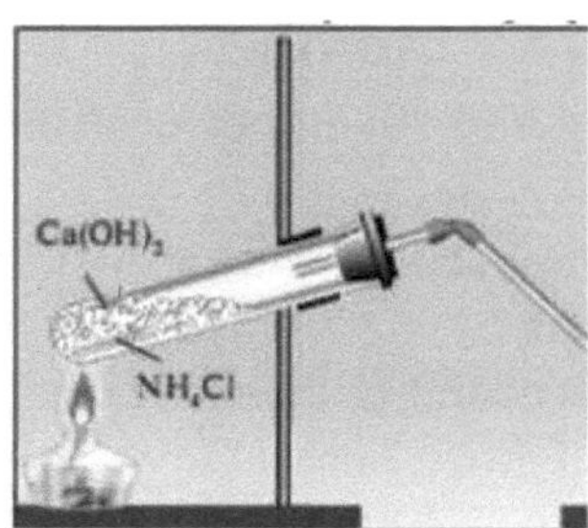

5. Which answer correctly states the name of the ship?

Date

Exercise 13
Determination of zinc, antimony, thallium and cadmium cations in the mineralisate.

Laboratory training plan

1. Determination of zinc cation in mineralisate .
2. of antimony in the mineralisation .
3. Arsenic cation in mineralisate .

Learning objective: to acquaint students with toxicological significance and methods of analysis of metal poisons, copper and zinc cations extracted from biological objects by mineralisation (decomposition) using sulphuric and nitric acids.

Expected outcomes during lab training:

cations that may be present in the mineralisation from the filtrate;

2 Determination of zinc and antimony cations that may be present in the mineralisation from the leachate;

Necessary equipment for laboratory classes: Mineralisate, filter paper, funnel, reagents, test tubes, demonstration materials, tape, paper.

Laboratory equipment: Room equipped with specialised laboratory equipment

.

1 - TITLE
Zinc dithionate reaction (preliminary reaction) .

CHALLENGE 2
Separation and extraction of zinc cation from other cations using diethyldithiocarbamate.

CHALLENGE 3
Analysis of antimony cations
Complexation reaction with malachite blue

1. Which class do the substances in the picture used in forensic chemical analysis belong to? (H_2SO_4 , NNO_3, HCl).

Date

Exercise 14

Determination of bismuth and aluminium cations in mineralizate

Laboratory training plan

1. Determination of zinc cation in mineralisate .
2. of antimony in the mineralisation .
3. Arsenic cation in mineralisate .

Learning objective: to acquaint students with toxicological significance and methods of analysis of metal poisons, copper and zinc cations extracted from biological objects by mineralisation (decomposition) using sulphuric and nitric acids.

Expected outcomes during lab training:

cations that may be present in the mineralisation from the filtrate;

2 Determination of zinc and antimony cations that may be present in the mineralisation from the leachate;

Necessary equipment for laboratory classes: Mineralisate, filter paper, funnel, reagents, test tubes, demonstration materials, tape, paper.

Laboratory **equipment**: with special laboratory instruments, furnished room.

CHALLENGE 1

Bismuth compounds

CHALLENGE 2

Study of bismuth mineralisation

CHALLENGE 3

Aluminium compounds

4. ^{What does} **C water** ^{mean} **in the formula**

$$K = \frac{C_{org}}{C_{suv}}$$

5. What does A stand for in the formula?

$$R = \frac{A}{N} \cdot 100\,\%$$

Date

24

Exercise 15

Determination of cobalt and tin cations in mineralisate.

Laboratory training plan

1. Cobalt from mineralisate cation determination.

2. Tin from mineralisate cation determination.

Objective of the exercise : metallic poisons cobalt and tin extracted from biological objects by mineralisation (decomposition) using sulphuric and nitric acids. to introduce students to the toxicological significance of cations and methods of analysis.

Expected outcomes during lab training:

1. Determination of the cobalt cation that may be present in the mineralisation from the leachate;

2 Determination of cobalt and tin cations that may be present in the mineralisation from the leachate;

Necessary equipment for laboratory classes:

Mineralisate, filter paper, funnel, reagents, test tubes , demonstration materials, tape, paper.

Laboratory **equipment**: with specialised laboratory instruments. furnished room

CHALLENGE 1

Cobalt compounds

CHALLENGE 2

Review of Mineral Resources for Cobalt

CHALLENGE 3

Tin compounds

Date

Exercise 16
arsenic content in mineralisate, analysis of an unknown object

Laboratory training plan

Arsenic cation in mineralisate .

Exercise will not last : To acquaint students with the toxicological significance and methods of analysis of metal poisons, zinc cations extracted from biological objects by mineralisation (decomposition) using sulphuric and nitric acids.

Expected outcomes during lab training:

cations that may be present in the mineralisation from the filtrate;

2 Determination of zinc and antimony cations that may be present in the mineralisation from the leachate;

Necessary equipment for laboratory classes: Mineralisate, filter paper, funnel, reagents, test tubes, demonstration materials, tape, paper.

Laboratory equipment: Room equipped with specialised laboratory equipment

.

CHALLENGE 1

Arsenic (margimush) <u>Applications and toxicological significance</u> .

CHALLENGE 2

Zang e r-Bl e kr e action margimush

CHALLENGE 3

Reaction of silver diethyldithiocarbamate with solution in pyridine .

1. The formula of which substance is shown in the figure?

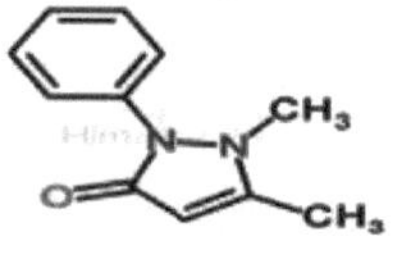

2. Which substance's formula is shown in the figure?

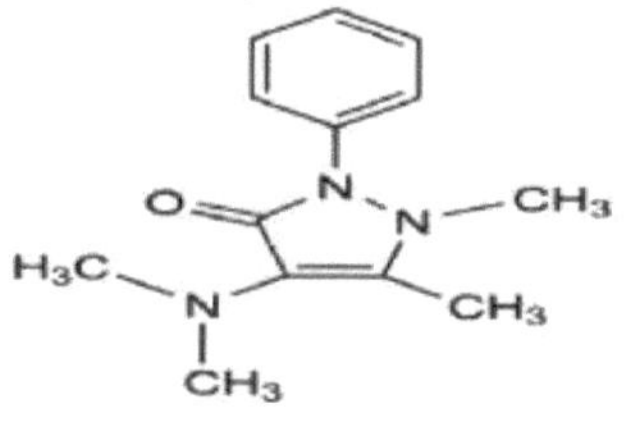

What is the formula of the substance shown in the figure?

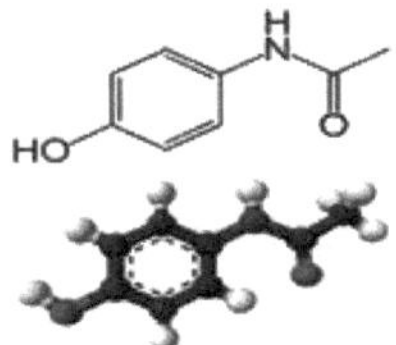

What is Analgin used for?

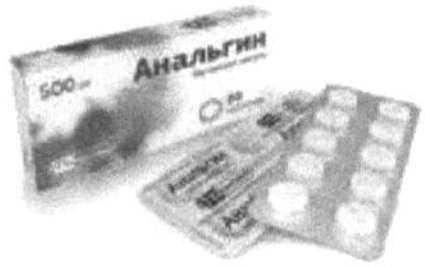

Date

Exercise 17
Liver and kidney damage. Determination of the quality and quantity of mercury by degradation.

Laboratory training plan

Determination of the quality and quantity of mercury

Exercise will not last: To introduce students to the toxicological significance of a toxic metal cation released from a biological object and methods of analysis.

Necessary equipment for laboratory classes: paper, funnel, reagents, test tubes, demonstration materials, tape, paper.

Laboratory equipment: Room equipped with specialised laboratory equipment
.

CHALLENGE 1

Destructive separation of mercury compounds in biological objects by concentrated sulphuric and nitric acids.

CHALLENGE 2

Structural analysis

CHALLENGE 3

The amount of mercury cation is determined from the composition of the resulting solution .

1. What plant is shown in the picture?

What formula is given?

$$m = C_{мин} * V_{мин} * 10^{6}$$

3. what is C min in the formula

$$m = C_{мин} * V_{мин} * 10^{6}$$

Date

Exercise 18
Forensic Chemistry. Forensic chemists, their duties, ethics and deontology. The writing and defence of expert reports.

Laboratory training plan

1. Forensic chemical report.

2. chemical and toxicological expertise.

Learning objective: To familiarise students with forensic chemical examination, its peculiarities, the plan of chemical-toxicological examination, duties of chemists-experts.

Expected outcomes during lab training:

1- There will be information about the forensic chemical examination, bureau, staff .

Forensic chemistry report. Chemistry will have information on toxicological examination. **Required equipment for laboratory work:** Destruct, filter paper, funnel, reagents, test tubes, demonstration materials, tape, paper.

Laboratory Equipment : A room equipped with specialised laboratory equipment.

CHALLENGE 1

Certificate of toxicological expertise writing guide.

CHALLENGE 2

Sample structure of an inspection report.

CHALLENGE 3

Statement of the act

1. What substance is shown in the picture?

O CH3
H3C N   N
O   N   N
CH3

what substance is depicted in the picture?

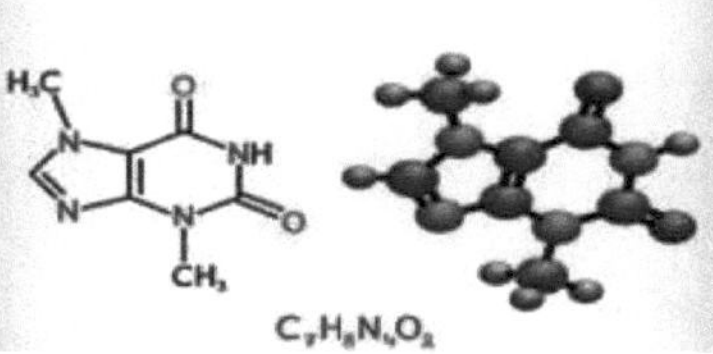

What substance is depicted in the figure?

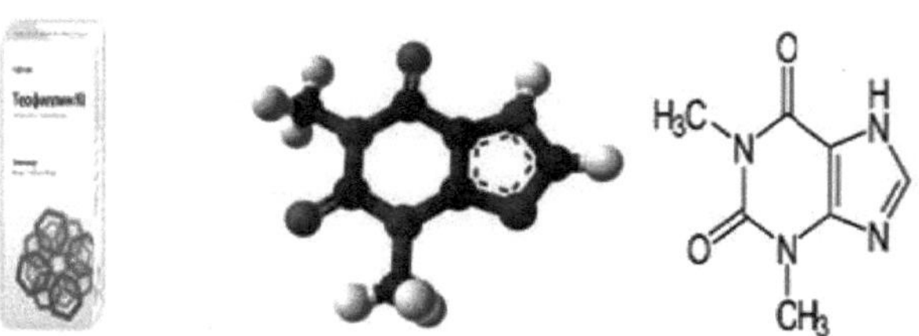

Exercise 19
Extraction of toxic substances from biological objects using polar solvents.

Extraction of toxic substances from biological objects using polar solvents. Separation methods using the acidified water method.

Practical training plan

1. Separation of acidic, neutral and weakly basic toxic substances using acidified water in the chemical-toxicological study of biological objects.

2. Extraction of the obtained extract with organic solvent
in acidic and alkaline environments, separating the toxic substance into two groups.

3. A study of aqueous extract of paxicarpine.

Topic: Separation of toxic and potent substances with acidic, neutral and weakly basic properties from the composition of a biological object by the acidified water method and explanation of the theoretical basis of the method.

Expected outcomes during the practical training:

To study the method of isolation of acidic, neutral and weakly basic toxic and potent substances using acidified water in chemical-toxicological studies of physical evidence .

2. Separation of "organic poisons" into two groups by extraction from the obtained extract with organic solvent (chloroform) in acidic and alkaline medium.

Comparison of general and particular methods for the separation of acidic, neutral and weakly basic substances by physical evidence. The choice of separation method is based on the nature of the object and the physical and chemical properties of the substance sought .

4. To study methods of purification of toxic substances from "foreign" substances isolated by extraction under acidic conditions using organic solvent.

Necessary equipment for practical classes: test tubes, separating funnels, reagents, demonstration materials.

Laboratory equipment : A room equipped with specialised laboratory equipment.

CHALLENGE 1

objects using water acidified with oxalic acid (A.A.Vasil and method) .

CHALLENGE 2

biological objects using water acidified with sulphuric acid solution (V.F. Kramarenko's method).

Stas-Otto method, its essence, advantages and disadvantages?

Date

Exercise 20
Liquid-liquid extraction. Determination of phenacetin, salicylic acid, acetylsalicylic acid in the extract.
extract.

Practical training plan

1. Toxicological significance of phenacetin, salicylic acid, acetylsalicylic acid.
2. Their methods of analysis.

Purpose of the training: to introduce students to the toxicological significance and methods of analysis of xanthine-containing alkaloids that pass into the organic solvent layer in an acidic environment.

Expected outcomes during the practical training:

During laboratory classes students study the conditions of qualitative reactions of alkaloids belonging to phenacetin and xanthine derivatives in pure solutions and in solutions of unknown composition, write down the results obtained in the workbook and confirm them with the teacher.

Necessary equipment for practical classes: test tubes, separating funnels, reagents, demonstration materials.

Laboratory equipment : A room equipped with specialised laboratory equipment.

CHALLENGE 1

Liquid-liquid extraction

CHALLENGE 2

Reactions for the detection of F e nacetyl tin Reaction for the formation of azo dye.

CHALLENGE 3

Determination of the proportion of salicylic acid

Date

Exercise 21
Determination of phenazone (antipyrine), propifenazone (amidopyrine), metamizole sodium (analgin), phenylbutazone (butadione) in secretions and liquid substances by the YuQX method.

Practical training plan

1. Toxicological significance of the substances phenazone (antipyrine), propiphenazone (amidopyrine), sodium metamisole (analgin), phenylbutazone (butadione).

2. Their methods of analysis.

Purpose of the lesson: To acquaint students with the toxicological significance and methods of analysis of phenazone (antipyrine), propifenazone (amidopyrine), sodium metamisole (analgin), phenylbutazone (butadione).

Expected outcomes during the practical training:

During laboratory sessions, students study the conditions of qualitative reactions of phenazone (antipyrine), propifenazone (amidopyrine), sodium metamisole (analgin), phenylbutazone (butadione) in pure solutions and in solutions of unknown composition.

Necessary equipment for practical classes: test tubes, separating funnels, reagents, demonstration materials.

Laboratory equipment : A room equipped with specialised laboratory equipment.

CHALLENGE 1

Determination of antipyrine from separators and liquid material (p e shab).
Actions to detect analgin

CHALLENGE 3

Determination of pyrazole derivatives by YuQX method

Date

<h1 style="text-align:center">Exercise 22</h1>

<h2 style="text-align:center">Preparation of reagents used in the analysis of alkaloids. Determination of indole alkaloids strychnine, brucine, reserpine in separators and liquid materials.</h2>

Practical training plan

1. Toxicological significance of strychnine and brucine alkaloids.
2. Their methods of analysis.

Topic: Students will be introduced to the toxicological significance of the substances of indole derivatives (strychnine and brucine), which are extracted with an organic solvent layer in an acidic medium, and the methods of their analysis.

Expected results during the internship:

During the internship, students learn the conditions for the determination of substances included in indole derivatives (strychnine and brucine) from pure solutions and unknown composition by qualitative reactions .

Necessary equipment for practical classes: test tubes, separating funnels, reagents, demonstration materials.

Laboratory equipment: A room equipped with specialised laboratory equipment.

<h2 style="text-align:center">CHALLENGE 1</h2>

Preparation of reagents used in the analysis of alkaloids .

CHALLENGE 2

Determination of strychnine reaction

CHALLENGE 3

Detection of brucine

Date

Exercise 23
Determination of purine alkaloids (caffeine, theobramine, theophylline) in extracts and liquid materials. Caffeine dependence. Spectrophotometric analysis.

Practical training plan

1. Toxicological significance of purine alkaloids (caffeine, theobromine, theophylline).

2. Their methods of analysis

Purpose of the lesson: to introduce students to the toxicological significance of weakly basic alkaloids passing into the organic solvent layer in acidic medium, as well as to the methods of analysis.

Expected outcomes during the practical training:

-Students study the conditions for the determination of qualitative reactions of alkaloids belonging to the derivatives of xanthine and indole in pure solutions and in solutions of unknown composition.

- Perform qualitative reactions for caffeine, theobromine, theophylline.

Necessary equipment for practical classes: test tubes, separating funnels, reagents, demonstration materials.

Laboratory equipment : A room equipped with specialised laboratory equipment.

CHALLENGE 1

Definition of coffee in a hotel _

CHALLENGE 2

Obromine detection is divided _

CHALLENGE 3

Spectrophotometric determination of purine alkaloids

Date

Exercise 24
Determination of nicotine, anabasine and paxicarpine alkaloids in extract and liquid material. Tobacco dependence. Microcrystalline analysis.

Practical training plan

1   Toxicological significance of nicotine, anabasine and paxicarpine alkaloids.

2   . Methods for analysing them.

Training objective: to familiarise students with the toxicological significance of nicotine, anabasine and paxicarpine alkaloids and methods of their analysis.

Expected outcomes during the practical training:

During practical classes students study the conditions of qualitative reactions of nicotine, anabasine and pachycarpine alkaloids in pure solutions and in solutions of unknown composition.

Necessary equipment for practical classes: test tubes, separating funnels, reagents, demonstration materials. **Laboratory equipment**: A room equipped with special laboratory equipment.

CHALLENGE 1

Quality of response to nicotine alkaloid

CHALLENGE 2

Quality actions for the alkaloid anabasine

CHALLENGE 3

Quality reaction to the alkaloid paxicarpine

Exercise 25
atropine, scopolamine, quinine and papaverine alkaloids in extracts and liquid materials. Application of gel chromatography.

Practical training plan

1   . Toxicological significance of tropane derivative alkaloids (atropine, hyoscyamine, scopolamine) and methods of their analysis.

2   . Toxicological significance and methods of analysis of the alkaloids comprising the quinoline derivative (xy n v).

Learning objective : Students will be taught the use of tropane-preserving alkaloids, cases of poisoning by them, metabolism, separation **from** physical evidence and methods of analysis.

Expected outcomes during the practical training:

students with methods of special separation of alkaloids related to tropane and quinoline derivatives, toxicological significance, methods of qualitative and quantitative analysis, as well as methods of determination of toxicological significance, quality and quantity of alkaloids related to tropane derivatives.

Necessary equipment for practical classes: test tubes, separating funnels, reagents, demonstration materials.

Laboratory equipment: Room equipped with specialised laboratory equipment

.

CHALLENGE 1
Reactions for the determination of atropine alkaloid

CHALLENGE 2
Determination of the reaction of the alkaloid scopolamine **to the alkaloid quinine**

Exercise 26

Determination of quinine and papaverine alkaloids in extracts and liquid materials. Application of gel chromatography .

Practical training plan

1 . Toxicological significance of tropane derivative alkaloids (atropine, hyoscyamine, scopolamine) and methods of their analysis.

2 . Toxicological significance and methods of analysis of the alkaloids comprising the quinoline derivative (xy n v).

Learning objective : Students will be taught the use of tropane-preserving alkaloids, cases of poisoning by them, metabolism, separation **from** physical evidence and methods of analysis.

Expected outcomes during the practical training:

students with methods of special separation of alkaloids related to tropane and quinoline derivatives, toxicological significance, methods of qualitative and quantitative analysis, as well as methods of determination of toxicological significance, quality and quantity of alkaloids related to tropane derivatives.

Necessary equipment for practical classes: test tubes, separating funnels, reagents, demonstration materials.

Laboratory equipment: A room equipped with specialised laboratory equipment.

CHALLENGE 1

Quinine alkaloid detection reactions

CHALLENGE 2

Determination of the reaction of the alkaloid papaverine

TASK 3 reactions for quinine alkaloid

Exercise 27

the substances caine, dykaene, dimedrol, lidocaine in secretions and liquid materials.

liquid materials. FEC method of analysis do _

Practical training plan

1.   Toxicological significance of novocaine, dicaine, dimedrol, lidocaine substances and methods of their analysis.

Learning objective: students will be taught the use of novocaine, dikaine, dimedrol, lidocaine, cases of poisoning, metabolism, separation from evidence, methods of analysis.

Expected outcomes during the practical training:

students with methods of special separation of novocaine, dikaine, dimedrol, lidocaine substances, their toxicological significance, methods of qualitative and quantitative analysis, as well as methods of toxicological significance, qualitative and quantitative determination of alkaloids included in tropane derivatives.

Necessary equipment for practical classes: test tubes, separating funnels, reagents, demonstration materials.

Laboratory Equipment : A room equipped with specialised laboratory equipment.

CHALLENGE 1

Novocaine detection reactions

CHALLENGE 2

novocaine from the p e shob

Date

40

Exercise 28

Determination of dimedrol, amitriptyline and depressants in secretions and liquid substances. Analysis by TDSIS

Practical training plan

Toxicological significance of antidepressants .

2. Methods for analysing them.

Training objective: to introduce students to the toxicological significance of **amitriptyline**, venflaxacin, mirtazapine, fluoxetine and methods of analysis.

Expected outcomes during the practical training:

During the internship, students learn the conditions for qualitative testing of amitriptyline, venflaxacin, mirtazapine, fluoxetine in pure solutions and in solutions of unknown composition .

Necessary equipment for practical classes: test tubes, separating funnels, reagents, demonstration materials.

Laboratory equipment: Room equipped with specialised laboratory equipment

.

CHALLENGE 1

TDSIS analysis of antidepressants

CHALLENGE 2

TDSIS The analysis proceeds in several stages:

CHALLENGE 3

Thermodesorption surface ionisation spectroscopic analysis of sertraline

Exercise 29
Diuretics in forensic toxicological practice
practice. Their analysis by High Screening

Laboratory lesson plan

1. analysis of drugs by YUQX method 2. Preparation of thin layer sorbent plates. Analysing the conditions of activity level enhancement and activity determination. Purpose of training: to teach students to use the YUQX method in the analysis of drugs in laboratory classes.

Expected outcomes during laboratory sessions:

Students will learn about the physical phenomena occurring in the process of thin layer chromatography, the application of the method in the purification of drugs from impurities, the determination of the quality of substances, its importance, types of thin layer chromatography. chromatographic analysis, Rf and Rfst indicators are presented, the conditions for its determination .

CHALLENGE 1

Drug identification by YUQX method

CHALLENGE 2

Diuretics :

Date

Exercise 30
Anti-inflammatory agents in forensic toxicology.
toxicology. Analyse them in the YuSSX method.

Laboratory lesson plan

1: Drug analysis by the USX method

Expected outcomes during laboratory sessions:

Students will learn about the physical phenomena occurring in the process of thin layer chromatography, the application of the method in the purification of drugs from impurities, the determination of the quality of substances, its importance, types of thin layer chromatography. chromatographic analysis, Rf and Rfst indicators are presented, the <u>conditions for its determination</u> .

TASK 1

Identification of *anti-inflammatory drugs in forensic toxicological practice*
CHALLENGE 2

YuSSX analysis

Date

Exercise 31
Antidiabetic drugs in forensic and in forensic toxicology. Analyse them by High Screening.

Laboratory lesson plan

1. drug analysis by the Yu method

Q X.

Expected outcomes during laboratory sessions:

Students will learn about the physical phenomena occurring in the process of thin layer chromatography, the application of the method in the purification of drugs from impurities, the determination of the quality of substances, its importance, types of thin layer chromatography. chromatographic analysis, Rf and Rfst indicators are presented, the <u>conditions for its determination</u>.

CHALLENGE 1

Antidiabetic drugs in forensic toxicology.

CHALLENGE 2

High-level analysis

CHALLENGE 3

antidiabetic drugs

Date

Exercise 32

Methane and carbon monoxide poisoning. Analysis of blood carboxyhaemoglobin by chemical and physical methods by chemical methods.

Practical training plan

1. Toxicological significance of carbon monoxide poisoning .

2. Analysis of blood carboxyhaemoglobin by chemical and physicochemical methods .

Training objective : to familiarise students with the toxicological significance of carboxyhaemoglobin and methods of **its** analysis.

Expected outcomes during the practical training:

During practical classes students learn the conditions of blood carboxyhaemoglobin analysis by chemical and physicochemical methods.

Necessary equipment for practical classes: test tubes, separating funnels, reagents, demonstration materials.

Laboratory Equipment : A room equipped with specialised laboratory equipment.

<u>Analyses of the condition are as follows:</u>

Chromatograph: Chromatograph Svet-110

Column: steel - polysorb-1, length 100 mm, inner diameter - 3 mm.

Column thermostat temperature $50\,^{0}C$, evaporator temperature $150\,^{0}C$.

Carrier gas: nitrogen, speed 30 ml/min.

Flame detector. Sensitivity $10*10^{10}$. Chromatograms are recorded on a KSP-4 potentiometer. 720 mm tape speed/well.

<u>Results of inspections and their analysis</u>

The identification of methane depends on the absolute dissolution time. Under the above conditions, the ignition time of methane is 3 minutes.

CHALLENGE 1

Methods of chemical production of carboxymoglobin (carbon dioxide) . Potassium (III) hexacyanate - K $_3$ ferrate [Fe(CN) $_6$] - reaction with red blood salt

TASK 2 *action with copper sulphate solution*

CHALLENGE 3

Reaction with caustic *sodium alkali solution*

CHALLENGE 4

Reaction with formaldehyde _ _ _
Reacting with presenters at e tati .

Exercise 33

Dialysis. Poisons that can be isolated from an object using water. Extraction and determination of mineral acids, alkalis and their salts from dialysate . Mineral acids (sulphuric, hydrochloric, nitric acids).

Practical training plan

1. Dialysis. Toxicological significance of poisons isolated from a facility using water. .

2. Methods of isolation and determination of mineral acids, alkalis and their salts from dialysate.

Learning objective: To introduce students to the toxicological significance of mineral acids, alkalis and their salts and methods of analysis.

Expected outcomes during the practical training:

During the internship, students learn the conditions of extraction and determination of mineral acids, alkalis and their salts from dialysate .

Necessary equipment for practical classes: test tubes, separating funnels, reagents, demonstration materials.

Laboratory Equipment : A room equipped with specialised laboratory equipment.

CHALLENGE 1

Dialysis of water-soluble toxic substances from the facility
extraction method

CHALLENGE 2

mineral acids in dialysate

TASK 3 *action with barium chloride*
Reaction with presenters _ _ _
Reaction with sodium rhodizonate .

CHALLENGE 4

Reaction with Dif and Nilamine.
Reaction with Brucine.
<u>*Campaign using woollen and cotton materials .*</u>

CHALLENGE 4

Checking dialysate for toxic substances in salt form
Sodium and potassium nitrites ($NaNO_2$, KNO_2)

Date

46

Exercise 34

Agricultural pesticides, organophosphorus pesticides, organochlorine pesticides. The isolation and determination of chlorophos, carbophos, hexachlorane.

Practical training plan

1. Toxicological significance of agricultural pesticides, organophosphorus, organochlorine pesticides .

2. Separation and determination of the substances phaseolon, hexachloran .

3. Solving a situational problem in practice

Learning Objective : Toxicological significance of agricultural pesticides, **organophosphorus**, organochlorine pesticides . and familiarise students with the methods of analysis.

Expected outcomes during the practical training:

During practical classes students study the conditions of qualitative reactions of substances phaseolon, hexachloran in pure solutions and in solutions of unknown composition.

Necessary equipment for practical classes: test tubes, separating funnels, reagents, demonstration materials.

Laboratory Equipment : A room equipped with specialised laboratory equipment.

CHALLENGE 1

Analysis of Carbophos
Extraction from the composition of a biological entity.

CHALLENGE 2

Reaction with diazotised sulphanilic acid
Whether an action is performed on the allocated asset

CHALLENGE 3

Chlorophos analysis
Extraction from the composition of a biological entity.

CHALLENGE 4

Fosalone analysis
Reaction of silver nitrate with alcohol solution

Date

47

Exercise 35

**Topic: Synthetic thyroid derivatives._ Cypermethrin, Danitol, De t cis.
Methods of their extraction from objects and analysis. Writing and defence
of expert reports.**

Practical training plan

1. toxicological significance of synthetic pyrethroid derivatives.

2. Cypermethrin, danitol, de t cis. methods for their extraction and analysis from
objects.

**Training Objective : To familiarise the students with the toxicological
significance of** synthetic pyrethroid derivatives and methods of analysis.

Expected outcomes during the practical training:

During practical exercises, students learn the conditions for qualitative reactions
of cypermethrin, danite and synthetic pyrethroids in pure solutions and in
solutions of unknown composition.

Necessary equipment for practical classes: test tubes, separating funnels,
reagents, demonstration materials. **Laboratory equipment**: A room equipped
with special laboratory equipment.

CHALLENGE 1

*Separation and analysis of synthetic pyrethroids from biological fluids and
objects*

Extraction from blood .

CHALLENGE 2

Extract from p e shob

Extraction from a biological entity .

CHALLENGE 3

GSX - analysis of synthetic *pyrethroids :*

CHALLENGE 4

UV - spectrophotometric analysis :

Date

Exercise 36
Topic: Substances that can be separated and analysed
by a separate method. Identification of an unknown object.

During the practicum, students will:

1. Determination of bromine from the composition of a biological object.
2. Determination of iodine from the composition of a biobject.
3. Determination of fluoride from the composition of a biobject.
4. Learn the work of identifying chlorine from the composition of a biological object.

Methodological instructions for practical classes

1 Determining the appearance of the item being inspected (colour, smell, presence or absence of foreign substances, preservatives, weight) and recording it in the workbook.

2. Determining the pH state of the object using a universal indicator and ionometer.

CHALLENGE 1
Determination of bromine from a biosample

CHALLENGE 2

Determination of iodine from the composition of a biological object
TASK 3 Determination of fluoride from a biological object Determination of fluoride from a biological object

FOR IMMEDIATE RELEASE

LIST OF REFERENCES USED

Primary Literature:

Ikromov L.T., Mirkhaitov I., Lozhiev M.A., Yuldashev /. O. Toxicological

chemistry. Textbook. Tashkent, Extremum Press. 2010. 593 c .

2. Vergejczyk T. ICS. _ Toxicological chemistry. Textbook. - M .: MED press inform, 2009. 400 c.

3. Vergeichik T.H. Toxicological chemistry. Textbook. - Moscow: MED Press Inform, 201 3.430 p.

4. Principles of forensic toxicology / edited by Barry Levine. 2nd. cr., ed. vow. updated, p. 401.

ADDITIONAL REFERENCES

5. President of the Republic of Uzbekistan 7 February 2017 "0 Uzbekistan On the strategy of actions for further development of the republic" Resolution№ PF-4947. 0 collection of legal documents of the Republic of Uzbekistan, 2017, no. 6, article 70

6. I.T. Ikramov and the textbook "Forensic Toxicology". Tashkent , 2020 _ _ 667 б .

7. Malkova G. G . Selected lectures on goxicological chemistry. Perm, 2005. - C. 87-99.

8. Toxicological Chemistry Nod. ed. irof.T.V.11ls1Snova-M: "Ya eotar Media", 2008.

9. toxicological chemistry module according to _ laboratory class exercises _ for methodical guidelines. Tashkent. 2020

INTERNET SITES:

www.ziyonet.uz www.astokscem./nu/ www.sudmed.ru

www.rc-sme.ru

Introduction to toxicological chemistry. Premises and laboratory equipment. Requirements for them.

Understanding physical evidence. Preliminary examination of physical evidence: pH-environment, colour, odour and preliminary examination of the object for the presence of individual substances (acids, alkalis, cyanides, white arsenic).

Removal of toxic substances from the site by steaming. Examination of first distillate for cyanidic acid

Analysis of volatile substances in the distillate for formaldehyde and acetone.

Analysis of acetic acid and phenol by volatile substances in the distillate

Examination of the second distillate for chloroform, chloral hydrate, carbon tetrachloride and their differentiation.

Chemical analysis of methyl, ethyl, amyl alcohols from distillate.

Introduction to gas-liquid chromatographs. Determination of the quality and quantity of alcohol in blood and urine by GSX method.

Wet mineralisation of the target in the presence of sulphuric and nitric acids. Denitrification of mineralisation and preparation for assays

Determination of barium and lead cations after separation of the precipitate from the mineralisation.

Determination of manganese, chromium, cations in mineralisate.

Determination of copper and silver cations in mineralisation

Determination of zinc, antimony, thallium and cadmium cations in mineralisation

Determination of bismuth and aluminium cations in mineralizate.

Determination of cobalt and tin cations in mineralisate.

Determination of arsenic cation in mineralisation . Analysis of an unknown object.

Liver and kidney damage. Determination of the quality and quantity of mercury by degradation.

Forensic Chemistry. Forensic chemists, their duties, ethics and deontology. Writing and defence of expert reports

Extraction of toxic substances from biological objects using polar solvents. Separation methods using the acidified water method.

Liquid-liquid extraction. Determination of phenacetin, salicylic acid, acetylsalicylic acid in the extract.

Phenazone (antipyrine), propiphenazone (amidopyrine), sodium metamizole (analgin), phenylbutazone (butadione) from secretions and fluids determination of substances by the YuQX method.

Preparation of reagents used in the analysis of alkaloids. Determination of indole alkaloids strychnine, brucine, reserpine in separators and liquid materials. Determination of purine alkaloids (caffeine, theobramine, theophylline) in extracts and liquid materials. Caffeine dependence. Spectrophotometric analysis. Determination of nicotine, anabasine and paxicarpine alkaloids in extract and liquid material. Tobacco dependence.

Microcrystallographic analysis.

Determination of atropine, hyoscyamine, scopolamine alkaloids in extracts and liquid materials. Application of gel chromatography

Determination of quinine and papaverine alkaloids in extracts and liquid materials. Application of gel chromatography.

Determination of novocaine, dicaine, dimedrol, lidocaine substances in secretions and fluid samples. FEC analysis.

Determination of dimedrol, amitriptyline and depressants in secretions and liquid substances. Analysis by TDSIS

Diuretic agents in forensic toxicological practice.

We analyse them using the YuQX screening method.

Anti-inflammatory agents in forensic toxicology. Analysing them in the YuSSX method.

Antidiabetic drugs in forensic toxicological practice. We analyse them by YuQX screening.

Methane and carbon monoxide poisoning. Analysis

blood carboxyhaemoglobin by chemical and physicochemical

Dialysis. Poisons that can be isolated from an object using water. Extraction of mineral acids, alkalis and their salts and [33] determination from dialysate. Mineral acids (sulphuric, hydrochloric, nitric acids).

Agricultural pesticides, organophosphorus, 34 organochlorine pesticides. Separation and determination of chlorophos, carbophos, hexachlorane substances.

Derivatives of synthetic pyrethroids. Cypermethrin, Danitol, 35 Decis. Methods of their extraction from objects and analysis. Writing and defence of expert reports.

Separately analysed substances. Identification of an unknown object. Writing and defence of expert reports.

List of references used

I want morebooks!

Buy your books fast and straightforward online - at one of world's fastest growing online book stores! Environmentally sound due to Print-on-Demand technologies.

Buy your books online at
www.morebooks.shop

Kaufen Sie Ihre Bücher schnell und unkompliziert online – auf einer der am schnellsten wachsenden Buchhandelsplattformen weltweit! Dank Print-On-Demand umwelt- und ressourcenschonend produzi ert.

Bücher schneller online kaufen
www.morebooks.shop

info@omniscriptum.com
www.omniscriptum.com

Printed by Books on Demand GmbH, Norderstedt / Germany